CeaC
(HOME)

D1462001

Guilt Free
& Green

A Year of
Eco-Friendly Ideas

Terri Paajanen
with
Madeleine Somerville
author of *All You Need is Less*

ISBN 978-1-63353-137-6

Out of all those millions and millions of planets floating around there in space, this is our planet, this is our little one, so we just got to be aware of it and take care of it.

- Paul McCartney

TABLE OF CONTENTS

FOREWORD

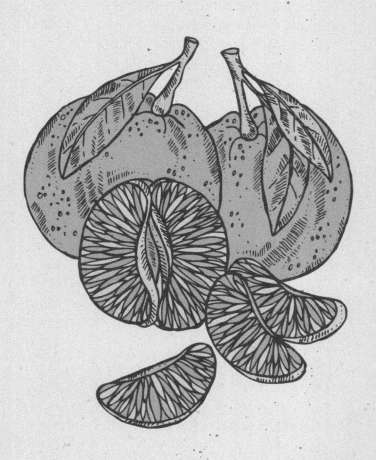

When taking on big environmental problems - climate change, overflowing landfills, relentless consumer culture or the floating garbage patch in the Pacific Ocean - it's tempting to go after big solutions, too.

This isn't necessarily wrong; large-scale government policy change and increased corporate accountability can be incredibly effective at mitigating our environmental impact on the world around us.

The problem is that big changes take time, and a lot of it.

Governments are infamous for moving at a glacial pace, which isn't helpfu when actual glaciers are melting before our eyes. Corporations often have to be bullied into cooperation by eco-friendly customers loudly demanding change. And in between campaigning for change and seeing it take place, we wait.

And wait. And wait. And wait.

A better solution is to take a dual-pronged approach - demand large-scale change from governments and corporations, sure, but while you're waiting for that to happen, make yourself useful! Create as much positive change as you can in your own life in the meantime. We don't have to wait for governments to tell us how to take on this problem or corporations to sell us a solution - we can empower ourselves.

Of course, figuring out how to get started is half the problem, especially if you're a gung-ho type like I am. Once I decide I want to do something I want to do it all, now.

Unfortunately, this unbridled enthusiasm often leads to me taking on too much, too soon, getting overwhelmed, burned out and fed up, then drowning my sorrows in metric tones of goat cheese.

Your comfort food may be different (fair-trade chocolate, perhaps?) but if your approach is the same, Guil Free & Green to XX is just what you (and I) need. It's a reminder that real change, sustainable change isn't made over the course of an impassioned day or a frantic week, but gradually, incrementally.

It's the first piece of advice I give everyone who wishes to embark upon a more Eco-friendly life: start small, start slow, and then keep going. Broken down into simple, easily implemented suggestions, the book you hold in your hand can help you change the world.

Contrary to popular belief, changing the world doesn't require drastic, life-altering changes like getting rid of your fridge or swearing off showers. Small, sensible shifts to your daily life are just as important and, when they add up, they can have a monumental impact. Tackle one suggestion each week - or even one each month, choose the pace suits you best!

An Eco-friendly life is never an all or nothing prospect, don't let the fact that you can't do everything prevent you from doing anything. This book contains dozens of ways to reduce your environmental impact, from buying in bulk to greening your mailbox; mixing up your own laundry soap or going shampoo-free.

All you need to do is begin. And now you know how. I wish you the best in your journey toward a simpler, more Eco-friendly life.

Namaste (just kidding)

Madeleine Somerville

GOING ORGANIC

GUILT FREE TIP

It probably needs no explanation why organic produce is the greener choice. Growing food and raising animals without huge doses of toxic chemicals is better for your body and better for the Earth.

One of the simplest places to start going green is to buy more organic goods, and getting to know what "organic" even means.

The rules for organic labeling are very specific, and aren't just a marketing ploy to possibly misrepresent a product. There are particular standards maintained at a national level by the United States Department of Agriculture (for American products at least). And these requirements must be followed, along with proof and paperwork, for a minimum of 3 years.

There is a fair bit of red tape involved in the standards, but a quick and suitable layman's summary is:

Organic crops must be raised without conventional pesticides, petroleum-based fertilizers or sewage-based fertilizer. Animals raised as organic must be fed only organic feed and have regular access to the outdoors. They are not to be given any antibiotics or growth hormones.

Basically, nothing toxic or artificial can be used in the growing of crops or the raising of animals.

Not Just Produce

Packaged and processed foods can be part of the organic world too, though it's a bit more complicated because there can be a dozen or more ingredients at hand in any one product.

Here's where you need to understand the labeling lingo. There are really 3 levels of organicness recognized by the USDA for legal labeling:

- "100% Organic" - this means just what it says, that every ingredient in the product is organic

- "Organic" - this means the product has at least 95% organic ingredients

- "Made with organic ingredients" - this label is for anything with 70% to 94% organic ingredients

Anything with less than 69% organic content doesn't get an official label. Labels like "all natural" or "eco-friendly" have no real meaning, so look for the official organic designation. Everything else is just marketing.

Not Just Food

Actually, you can find organic products outside the food industry altogether, which sometimes takes people by surprise. Many fabrics can also fall into this field, and you can get organic cotton and other fibers in clothes and bedding.

MAKE YOUR OWN
NON-TOXIC CLEANERS

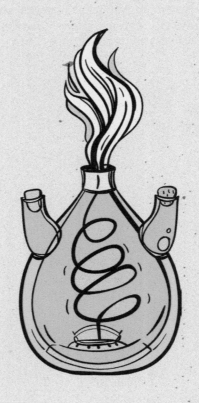

GUILT FREE TIP

A few drops of your favorite essential oil can give this cleaner any scent you like.

Thought about cleaning greener, but got lost with all the various cleaners on the market? Not sure how environmentally-friendly they are, or not happy with the higher price tags? Well, with a few handy recipes, you can make some excellent non-toxic cleaning products for around the house that are much cheaper than anything you can buy.

General Surface Cleaner

No matter what the commercials say, you don't really need to have a whole cabinet filled with cleaners for every specific purpose. One good general cleaner will go a long way, and you can make an easy one with basic ingredients.

- 1/2 cup white vinegar

- 1/4 cup baking soda

- 2 quarts water

Just mix everything together, let any fizzing settle down and store in a spray bottle. You can use this mix for lots of surface cleaning around the house. The vinegar is a naturally acidic disinfectant, and the baking soda gives just a hint of abrasive power. The baking soda can settle out if the bottle has been sitting for a while, so just give it a shake when those cleaning chores crop up.

Tub or Tile Cleaner

If you need a more robust cleaning solution for the tub, or even around the stove top in the kitchen, you just need to use a damp sponge and a sprinkling of baking soda. Just because it's a tougher job, doesn't mean you need harsher chemicals.

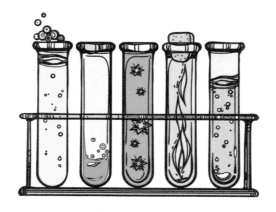

Toilet Bowl Cleaner

You can use the surface cleaner to give your toilet a basic wipe-down, but for scrubbing inside the bowl, you'll need a different recipe.

- 1/4 cup baking soda

- 1 cup white vinegar

Just dump both ingredients straight into the bowl and let it sit. The fizzing will help loosen the grime, and then you can use a standard toilet brush to finish the job.

Laundry Soap

For the more industrious DIY-er, you can even make your own laundry detergent.

- 1 bar of Ivory, Fels-Naptha ® or other pure soap - grated finely

- 1 cup Borax

- 1 cup washing soda (NOT baking soda)

Mix everything together and store in an air-tight container. Use just 1 tablespoon for an average load of laundry.

GET SOME FRESH AIR

GUILT FREE TIP

You know all that fluff you pull out of the dryer lint trap every week? That's tiny bits of your clothes, worn away by the dryer action. A clothesline doesn't do that, and your clothes will actually last longer.

Use all that free fresh air out there to reduce your electricity use on laundry day. It's not as quick or convenient as that handy dryer in the laundry room, but it really isn't that much work once you get into the routine of putting your clothes out on a line. Here's how to get started.

The Simple Approach

A folding rack would work both inside and out if you only have a small load to dry. Otherwise, you can go with one of the free-standing umbrella style racks for more space. You may need to sink a hole to anchor it down, depending on the model. They work fine (and are especially suitable for little yards), but the clothes aren't as spread out and it can take longer for your clothes to dry. Still, it's a fine option.

Or you can just get some sturdy rope and string up a stationary line between two points. You'll have to walk along the line as you put up the clothes since the line doesn't move.

A classic Clothesline

For the full clothesline experience, you might want to put up the standard type of line that moves by using pulleys at either end. This allows you to stand at one end, and move the line along as you add clothes. You can usually buy kits that contain a good length of proper vinyl-coated line, two pulleys and a connector.

Attach a pulley at either end of your intended clothesline, then loop the line through them both. You'll want to cut the length long enough to cover the distance twice with at least 6 inches extra for the connector. Once the connector twists the ends together, it should be fairly taut. Now you just need a non-rainy day to put out the clothes.

How long it takes will really depend on your clothes and the weather. A warm day with a light breeze can dry thinner fabrics (perhaps cotton sheets) in as little as an hour. On the other hand, a heavy pair of jeans can take all afternoon. On average, you should be able to dry 2 loads of laundry in a day though. Just make sure to wash clothes on sunny days if you can.

So how green is this tip? Well, on average, your electric dryer is going to create 4 1/2 pounds of carbon pollution for each and every load. If you air-dry 2 loads per week, that's 36 lbs. of carbon you're keeping out of the atmosphere in just one month. In terms of electricity use, a pretty standard dryer uses 5400W making it one of the biggest energy drains in the house.

THE NO POO MOVEMENT

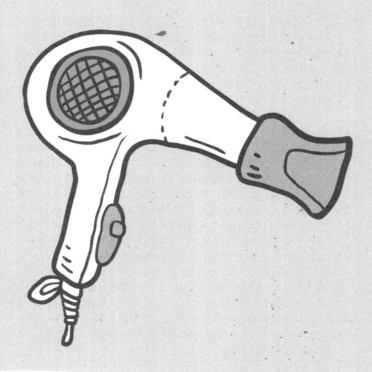

GUILT FREE TIP

It can take up to a month or more to see an improvement, so add a few hats or scarves to your wardrobe while you're transitioning.

I'll bet the title of this tip has you thinking, but probably about the wrong thing. This ever-so-clever expression is actually talking about shampoo, or rather the lack of it. Some people have stopped washing their hair with shampoo, and you might want to see the reasons.

The premise is that modern shampoos are really much too harsh on our hair, stripping away natural oils. Your body overcompensates by producing more oil, and so you need to wash more often. Not a sensible approach. Besides that, most commercial shampoos are loaded with chemicals that aren't great for the environment. Two of the worst are sodium lauryl sulfate and sodium laureth sulfate. They have been potentially linked to cancer (particularly the laureth form) and both are actually harmful to your skin. The paraben family of chemicals are also found in shampoos, and they have also been tied to cancers (cite). They're all bad for your body, and bad for the environment once they go down the drain. So how do you actually do this?

Not everyone handles their hair hygiene the same way, so there isn't a single "no poo" process. Generally, hair is given a brisk scrub with just water during your regular shower, which is enough to get rid of most dirt or dust. Some people leave it at that, especially if they have short hair. Others have added some natural "products" to their routine.

A little baking soda can help absorb oil and add some abrasive to give your hair a thorough cleaning without the host of chemical products. Just sprinkle a little in your hair, and use it to gently scrub your scalp. Rinse out well. And speaking of rinsing, an excellent conditioner for no-poo hair is apple cider vinegar. A quick splash worked through the hair is all you need, again make sure to rinse well. If you try going without shampoo, be prepared for a few weeks of adjustment. Your scalp is used to producing a lot of oil and it's not going to just quit that overnight. Eventually, things balance out and you can have naturally soft and non-greasy hair without the shampoo.

GET AROUND WITH PUBLIC TRANSPORT

GUILT FREE TIP

Electric buses are even cleaner than conventional ones, since they don't pollute at all. Your city doesn't have any electric buses in their transit fleet? Maybe it's time to write a letter about that.

*D*o you take your own car everywhere you go, loving the comfort and convenience of having your own vehicle? Well, today you should examine your transport habits and see if some changes can be made.

The greenest option is to either walk or take a bike. Unfortunately, there are lots of reasons and situations why that doesn't work for everyone. Your next choice is to go for public transportation. While bigger cities often have subways, trains or light rail options, the more common option is going to be the bus.

I'll admit that there is some controversy about whether or not a bus is actually the more eco-friendly choice (cite). Various studies have shown various things, and it mainly depends on how many people are in the bus. A full bus can represent several dozen cars not being driven, but one that is empty is clearly not helping the planet at all. I'm not simply suggesting you take the bus (or other form of transport) because its greener, I'm suggesting it because it will become greener as more people do it.

So how does one become a bus guru? The first step is to get a route map and learn how the system works. Some places have a flat-fare and some base the charge on the distance you're riding. Do your buses accept coins or must you carry tickets or a pass? What do the color-coded routes mean? Know the details and you're off to a good start. Carry a map with you until you get familiar with your local routes. Allow extra time in case you miss a connection.

Discouraged by how long it's going to take you to reach your destination? Remember that you're not doing the driving, so you can do all sorts of things during that time. Use a mobile device to do online work/play/whatever, read a book, or even knit. Maybe you'll make some friends on your regular route, and you can sit and chat. Don't consider it wasted time.

What about the costs? Again, that's a tricky question. Paying a per-trip fair is going to be the most expensive option, but frequent bus travel gets pretty cheap once you buy a pass. Comparing it to driving is difficult because you still may be making payments on a car whether you drive it or not. Given the cost of gas, it's definitely cheaper to take the bus providing you're not paying for a car in the background too.

BUY IN BULK

GUILT FREE TIP

Buying only the amounts you need will help reduce further waste. If you only need a tablespoon of something, having to buy an 8oz package is just silly.

This is a tried-and-true tip to help reduce all the paper and plastic waste that goes into food packaging. We're referring to bulk food stores where you scoop and bag your own. The simple thin plastic bags from these stores are recyclable and are a much better choice than the mixes of plastic, metal, foil and paper packages that so many food products are sold in. Even when you're dealing with recyclable materials, less is always better.

What can you buy at a bulk food store? My goodness, it might be easier to list things they don't carry. A good-sized store should have:

- Baking ingredients - flour, sugar, salt, chocolate chips, baking soda etc.

- Breakfast cereals, Granola, Pasta

- Beans, Soup mixes

- Cookies - brand names and generic

- Coffee - beans or ground

- Tea - loose or bagged

- Spices - ground or whole

- Pet food and treats,

- Dried fruit

- Nuts - roasted, raw, salted, flavored

- Candy - wrapped, loose, hard, gummy and anything else

- Liquid products - honey, syrups, nut butters

Many good bulk shops also have several organic lines of products so you can save on the garbage as well as buy organic. You'll probably see plenty of cost savings too, but not everything you buy in bulk is going to be cheaper than the packaged versions.

Store your bulk purchases in their original bags, or get a collection of jars or plastic containers. Dry goods like these are great for shelf storage without any special care. Also, if your store offers product information slips, take them if you can. It's very easy to buy 5 or 6 different kinds of beans on impulse, then forget how to cook them all.

Buying only the amounts you need will help reduce further waste. If you only need a tablespoon of something, having to buy an 8oz package is just silly.

BECOME A LOCALVORE

No chemicals

GUILT FREE TIP

Learn how to freeze or can your produce at home. That way, you can buy locally and in season, but save your fruits and vegetables for later in the year.

In this world of conscious eaters who make choices to avoid carbs, gluten or meat, the term localvore is a fairly new one that is part of the bigger movement to live more greenly. It's about making food choices to eat more locally grown and produced food as a more environmentally friendly decision.

We get so caught up in the chemicals involved in the actual growing or processing of food, we forget some of the "behind the scenes" pollution problems. In this case, it's all about the transportation. Are you happily choosing an organic mango, thinking that you're making the right choice? What about the pollution spewed out by the trucks, cargo planes or transport ships that went into bringing that exotic fruit to your local supermarket? Don't forget to factor in the energy used to keep perishables refrigerated for days (if not weeks) during their travels. According to David Suzuki, the average meal travels 1200km between farm and your plate (cite). That's roughly the distance between New York City and Chicago, and that's just for one single meal.

Going local means you're reducing the exhaust and fuel use that goes into moving food around the globe. Unfortunately, that's still only part of the whole picture. A local piece of fruit that was grown with a load of pesticides and weed-killer can still be worse than an organic piece that had to travel. You can't discount both ends of the equation. So when you do start buying local, go for organic or naturally grown food to really make it count.

The easiest way to get local is the farmer's market. It's fun, friendly and the best place for healthy local food. Also, watch the signs at the regular supermarket. Most will say where their produce has come from, so you can choose your strawberries from within your region rather than those that come from across the country. A side effect of eating locally usually means you have to start eating in season as well, though that really depends on where you live. Get in tune with the seasons, and you'll come to appreciate the changes in availability as the year shift from spring to summer to fall.

SOMETHING OLD, NOTHING NEW

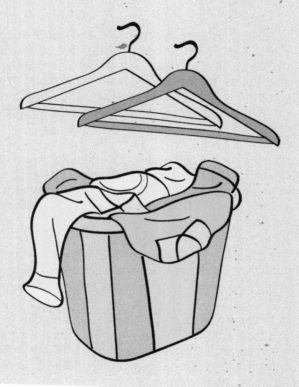

GUILT FREE TIP

Try to donate as much as you can from your household to thrift shops, so that other people can avoid buying new too.

*C*heap goods are everywhere, and we've become a society that throws things out in favor of something new without a moment's thought. It's wrong for so many reasons! Not only are we burning through our precious resources at a break-neck pace, we're going to soon be drowning in garbage and waste.

According to the EPA, the USA creates more than 250 million tons of garbage in a year, making each person responsible for about 4 1/2 pounds of trash each day (cite). It's time to put an end to it by reducing consumption of "stuff".

Break the cycle in your household by not buying new things. Try a challenge and for a week or month buying nothing new (other than food and hygiene products).

The first step is to stop needing new things in the first place. Make do with what you have and a lot of your purchasing will stop. Simply being tired of something isn't a great reason to replace it. As long as an item still works and has a purpose, it should be fine to keep using. Mending and small repairs can really help, and that will give you a chance to pick up new skills.

Even as you start to have more appreciation for your existing stuff, there is always going to be some sort of need as things truly wear out, break down or otherwise need replacing. But do you need to buy something "new" is the question. Can you replace that item with something second-hand? Many items in thrift shops are barely used. Used goods help reduce bulky trash in our landfills, and eliminate the need to use resources to make unnecessary new things.

Do you need something temporarily? Ask your friends and neighbors if one is available to borrow. Hardware and outdoor equipment can often be rented from large home improvement stores, meaning you don't have to clutter up your garage by buying stuff you'll only use once or twice.

Another new trend is the freecycle, where people with unneeded things post to various websites just to give them away for free. It saves them the hassle of trying to dispose of stuff, and it means someone else can get use out of it. It's green and free. What a combination!

Personally, I once got a huge bag filled with practically brand-new clothes for my daughter through a freecycle exchange, and it took me 5 years to finally work through all the sizes. The final pair of jeans is currently in her closet. And when she outgrew each item, it all went right back to a second-hand store to find yet another home.

ECO-FRIENDLY APPS

GUILT FREE TIP

Be aware that platforms and operating systems change frequently, so not all of these may be available or compatible with your exact phone or tablet.

*Y*es, technology can be eco-friendly, or at least it can be used as a tool for making more environmentally sound choices. That includes a collection of apps to put some eco-tools at your fingertips.

Think Dirty - If you're trying to green up your bath and beauty routines, this is a must-have app. You can use it to scan the bar code of your various personal care products, and it will tell you about any unpleasant ingredients that you might want to avoid. Their database includes more than 2,900 brands and over 200,000 individual products so you should be covered no matter what you're shopping for.

Oroeco - This is one of those apps that uses a social networking framework to add a competitive edge to your new lifestyle choices. It keeps track of your choices to let you know how much money you're saving, carbon you're emitting and more. Then you can mark your progress against friends in the network to motivate you to do more.

Farmstand - Not sure where to shop for locally grown produce? Farmstand has a database of nearly 9,000 markets around the world. Find ones closest to you, check on operating times and even browse photos submitted by visitors. It's a localvore's dream.

PaperKarma - Even in this digital age, we often find our physical mailboxes jammed full of wasteful junk mail. This is a neat app to get it all under control. All you need to do is photograph the offending junk, and the app scans it, recognizes the item and contacts the sender to remove you from their mailing list. It only works for mail that is actually addressed to you though.

Joulebug - Like Oroeco, this is a social-networking sort of app to help you make green choices every day. It's loaded with creative eco-suggestions to get you thinking about how you do things, and as you take on their challenges, you earn badges to brag about on your profile.

TRASH FOR TIGERS

GUILT FREE TIP

Get creative and find new ways to help the environment that work for your situation.

I thought I would share a unique project we do at our house, as an example of how you can tailor a green experience to your own lives.

We live in a rural area, and finally got tired of seeing trash along the road as we drove to and from home. Much of it is recyclable too, with aluminum cans and water bottles. So, my daughter and I started taking walks along the road with a bag to pick up this stuff, with the intention of adding it to our own blue bins each week. After a trip or two, we found a lot of beer cans and bottles. In our area, these can be returned to the store to collect the deposit. All of a sudden, our little clean-up walks had a financial element and Trash for Tigers was born.

Each beer can or bottle earns us a dime. I also decided to donate the proceeds from any wine bottles we accumulate at home, as they're a bigger earner at 20 cents apiece. So we started to turn in our finds, and we racked in the cash. After some family discussion, the plan was hatched that we would use this money to "adopt" endangered animals through the WWF (link). Every $50 we collected went to adopt an animal, providing funds to the WWF to help with conservation and education. So far we have a tiger and a jaguar, and the Trash for Tigers jar is getting to be about half full again. We've been talking about doing something different with our next donation, possibly a symbolic adoption of a few acres of conservation land through the Nature Conservancy (http://www.nature.org/).

So don't limit yourself to the usual tips (though they're an excellent way to build a green foundation in your home). Around here, we've cleaned up our neighborhood, donated money to a worthy cause, and also started building some global eco-awareness in our daughter. The exercise and fresh air from our walks hasn't hurt either.

CHOOSE CLOTH OVER DISPOSABLES

GUILT FREE TIP

More water is wasted in the production of disposables than you'll be using in your laundry, and it still doesn't take all the garbage into account.

We're talking diapers here, which may not be all that applicable to every family reading this book. But it's such a huge green tip that it can't be ignored.

Many people are waking up the trash problem caused by our disposable culture, and reducing the number of throwaway products that come into their homes. Diapers are a bit of a holdout mostly because the ease of getting rid of them as well as their contents in a matter of seconds is pretty appealing. If you truly want to be a green parent, look past the "gross" and consider the trashy reality behind disposable diapers.

Typically, between birth and complete toilet training, a child will produce 8,000 diapers. Or in other terms, the USA throws away 3.7 million tons of diapers each year (cite). However you look at it, that's a lot of garbage. For the complete picture, don't forget that a lot of resources and water goes into the production of disposable diapers, resources that would be better used elsewhere.

So how to get started with cloth? First, you need to get yourself supplied. I'll warn you up front that there might be a little sticker-shock when you see that a good set of cloth diapers can be several hundred dollars. Just remember that over the course of 3+ years, disposables will cost a few thousand.

You can go with cotton, hemp or bamboo. If all you care about is reducing your trash, then any of them will do. If you want to think a little broader about the resources used in creating the diapers then try to go with either hemp or bamboo. Cotton uses a lot of water in growth and production, and it's a heavily pesticide-laden crop unless you go organic.

A "full set" is about 24 diapers, though the specifics can vary depending on the type you get. You might also need to get waterproof diaper covers as well, but some all-in-one diapers have them built-in. Most modern cloth diapers are easy to use with Velcro or snaps and elastic legs, so don't let your fear of safety pins stop you from switching to cloth. They go on and off with hardly any effort.

I also highly recommend diaper liners along with your actual diaper supply. They are fine paper sheets (kind of like a used dryer softener sheets) that are laid in each diaper as you put them on. Solid waste is much easier to clean up if you can just lift out the liner at changing time. After that, you just toss them in the washing machine and then should clean up without any further effort.

And don't be swayed by the diaper companies who try to confuse the issue by claiming cloth diapers aren't really that green because you use so much extra water doing more laundry. It's extremely false.

DON'T BE E-WASTEFUL

GUILT FREE TIP

Think twice when you're oogling that new phone at the store. Do you really need it?

E-waste isn't just a form of Internet junk email, it's the whole new range of garbage we're producing from all the electronic devices being disposed of daily. Phones, computers, TVs, tablets and all the other various devices we're getting so dependent on. These items are too complex for simple recycling, and they're full of plastics and a wide range of metals. Not something that should be just dumped. Unfortunately, many people do. Don't be one of them.

use Less to start with

First of all, don't fall for the industry hype that you need a new phone or that you need a bigger TV. Unless there is something significantly wrong or lacking with your current device, be happy with it. Remember that each new model of smart phone represents more toxic trash as well as a waste of more resources.

I am as high-tech as they come, and yet I don't own a smart phone. I have an 8-year old phone that does little more than make phone calls. I keep a minimum emergency plan on it, costing $100 for the entire year. Basically for car problems, like when I locked the keys in the car 2 days ago. Would a fancy Internet-ready phone be handy? Most certainly. But with a brutally honest eye, I can say that I am home often enough to just use my regular computer for those things. And so, I remain basically phone-less.

Recycle Right

It's easy to overlook the need to recycle a tiny little cell phone because it seems insignificant, especially if you don't really know what to do with it. It's actually quite easy to recycle, even though these aren't things you can toss in the blue bin (yet).

Check with your local computer or business supply store. Many larger chains will take back related products, whether or not you purchased them there. If that doesn't work, you can also ask at your local hazardous waste disposal depot. They may take electronics themselves, or know who else would. You can also see about donating items that still have life in them. Computers and other devices are often refurbished and supplied to schools or low-income families.

How big a problem is this?

Well, the EPA says that there were nearly 2 million tons of electronic waste discarded in 2013 (cite). It's not just about the quantities either. There are some seriously toxic components in many electronics, such as lead, mercury, cadmium and gold. These aren't elements you want leaching into the soil or water table.

PLANT A TREE,
OR TWO

GUILT FREE TIP

Even if you only plant one new tree, it will produce 260 pounds of oxygen every year once it's matured.

This isn't an activity that has to be left only for Earth Day. Why not get into the habit of planting a few trees more often, and help out the planet with just a little effort. They help clean the air of excess carbon dioxide and provide us with fresh oxygen. They also create natural ecosystems for all kinds of birds, insects and wildlife too. I doubt you need convincing that trees are important. We're losing around 50 thousand square miles of forest every year, mostly due to clear-cutting for livestock pasture, so the planet can use all the new trees it can get.

How to Get it Done

Tree seedlings are usually available at any large nursery, though you might have to be clear on what you're after. Many people are looking for shrubs or strictly decorative plants. If you just want plain native trees, make sure that's what you're buying. Local environmental agencies sometimes have seedlings available at a lower cost, to encourage tree planting.

Then figure out where you want to plant your trees. Your own yard is a place to start, providing you have the space for it. Remember that a tree will be big after 5 to 10 years, with a crown taking up hundreds of square feet.

Once you've planted trees at home, look elsewhere. Offer to plant a few trees for your neighbors, or maybe on the grounds of the park, school or library. Get creative! Always ask first if you're thinking about putting trees on private land though. Abandoned places can sometimes use a tree, but you can never be sure that an empty lot won't be razed later for building. Stick to places that are more likely to stay stable.

Seedlings should be planted in late spring, when the weather is warm enough that your little tree can thrive and get settled for a few months before having to survive its first winter. If you can, visit your new tree every few weeks and give it a drink if you're having dry weather.

THE ENVIRONMENTALLY-
FRIENDLY MAILBOX

GUILT FREE TIP

*For a week, take note of your junk mail and piece
by piece, try to take the right steps to shift it to an
electronic format or stop it altogether.*

*F*or this tip, we're talking about the physical mailbox that the mail carrier visits every day, rather than your online email inbox. That's coming up in another tip.

Green up the Mail you want

Not all paper mail is truly junk, even though it might be wasting a lot of resources to get to your door. Do you really need all that mail to be in physical format? Most major stores offer their weekly flyers (and all the lovely sales) in email format now, you just need to subscribe. They are usually in a graphic layout so it's really just like flipping through the real flyer, just without the acres of paper. The same can be said for many catalogs too. Visiting websites can give you as much insight into someone's inventory as a catalog, maybe even more so since it can be updated whenever the store needs to. Statements from banks and utilities are another place you can trim your mail load. Most places will issue e-statements now, so you can get your bills or statements via email. Now that doesn't mean you have to pay your bills through online methods, just that your bill comes to your email.

Ditch the Mail you Don't

Now what about all the bushels of paper waste that you pitch in the trash right away because you don't want it, and never wanted it? Unfortunately, it can be tough to stem the tide. A note on your mailbox stating "no junk mail" can help, but that would only apply to un-addressed mail (like sale flyers). A mail carrier will deliver anything addressed specifically to you no matter what it is. Next time you get junk, get in touch with the company and ask to have your name removed from their list. If that doesn't work, draw a line through your address on the outside, with "Return to Sender" in bold letters. The post office will return it, and that should trigger your removal as a bad address.

Also, take care when you sign up for things in the first place. The more companies that have your address, the more junk mail will eventually find its way to your house.

TUCK IN YOUR
WATER HEATER

Boiling point

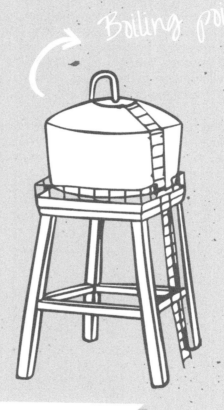

GUILT FREE TIP

If you're really interested in reducing energy loss with your hot water, you might want to consider an on-demand system to eliminate the tank altogether.

Insulation helps keep your precious heat inside the house, but some spots are going to need more coverage than others. For the moment, we're talking about the hot water tank.

Most modern hot water tanks are very well insulated and won't need too much attention. Just lay your hands on the sides. If you feel even the slightest warmth, then you want to add some insulation. You're using fuel or electricity to heat that water and don't want the heat to trickle away. On average, about 10 to 20% of a household power usage goes to keep the water hot.

Hardware stores sell foil and fiberglass blankets designed to wrap a hot water tank, and they are a great option to keep that heat in. They install easily with just heavy-duty tape. Before you buy a kit, take note of the size of your tank, the kits are sold to size.

After the tank, you need to check on the pipes. The pipes that run from the tank to the various points through the house will also leak heat when there is water sitting in them (which is most of the time). Foam tubes can be cut down to length with sturdy scissors and just fit the tubes over the pipes. Assuming you can get at the pipes, you can do this entire chore in an hour or less.

ON - DEMAND
HOT WATER

Heat water very quickly

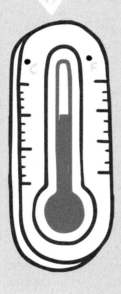

GUILT FREE TIP

If the cost is out of your budget, you can still help reduce your hot water energy usage with the insulation ideas in the last tip.

After reading our last tip, you might be doing some thinking about all the energy you're using just to keep that huge tank of water hot all through the day. Is that really the most effective or efficient way to have a hot water supply at home. There is an option.

An on-demand system gets rid of the tank completely, so there is no power used to maintain all that water when nobody needs it. Water is heated very quickly only when you turn on the tap. Not only do you save a considerable amount of energy this way, you can reclaim a big chunk of space in your basement by taking out that hulking tank.

A tank-less water heater usually operates on your whole house water supply, but you can get smaller ones to install right under the sink. These smaller ones can be very handy if your house is laid out in such a way that you have a sink at the other end of the home and would rather not have your hot water have to travel so long.

These heaters aren't a budget choice as they are more expensive than the usual tank arrangement. On the plus side, they do reduce your energy bill and a good on-demand heater will last twice as long as a standard tank. You have to consider the bigger picture to figure out the true costs. On-demand water heaters can be fueled by electricity or a gas, such as propane or natural gas.

TRY A NATURAL SWIMMING POOL

GUILT FREE TIP

Chlorinated water doesn't just stay in the pool. It gets into your soil every time you splash.

I'll admit right at the start that this isn't a bargain tip, or one that the average person is going to rush out an implement. On the other hand, it can be a very important thing to think about if you do happen to be considering putting in a swimming pool.

The conventional process for maintaining a swimming pool involves heavy doses of chlorine and other chemicals to keep the water free of bacteria and algae. If you want the joy of a swimming pool without the toxic soup, there is another way. There is a new trend of natural swimming pools you should learn more about. It's a lot like creating a natural pond ecosystem in your yard, to keep the pool clean. You'll have to do a bit more research than what we can explain here but the general approach is to have separate compartments built into the pool, connected so that the water flows evenly around the entire system. Then one part is filled with some kind of substrate (like gravel) and planted with a number of water plants. The plants and the gravel create a natural filtering system to keep the water clean, and it looks gorgeous too.

Not only will you be helping the environment by not maintaining all that chlorine in your yard, the water will be much safer and healthier for your body too. No more smelly residue on your skin and no more burning eyes after a swim. Since there is a large living component to a natural swimming pool, you'll have to keep your climate in mind when you design it. In a place with winters below freezing, you might have to replace the plants each year in the spring. It's something to discuss with your pool installation team.

And if you're wondering what the problem is, since your chlorinated water is safely inside the pool, how is it harming the environment? Actually, the main problem is that chlorine evaporates, adding to our air pollution issues. That's the main reason pool owners are constantly adding more to the water, sometimes even daily when the weather is warm. Don't be fooled by the serene look of a pretty blue pool.

LESS WATER
DOWN THE DRAIN

GUILT FREE TIP

Get in the habit! If you reflexively always do a full flush, then you won't be saving any water at all.

There haven't been too many creative innovations in the toilet industry since indoor plumbing was invented. A tank of water is released when you flush, to wash the contents of the bowl down away and out of your life. The problem is that you don't always need a full tankful of water. In fact, most of the time you need a whole lot less. And so, you're flushing clean water down the drain for no real reason. Time to improve your toilet game. Enter the dual-flush toilet.

They've been around since the 80s but are only now really getting to be popular and easy to find in North America. The idea is that there are 2 flushing levels, and you can choose whatever one you need for that moment. A light flush is fine for liquid deposits, but you can do the full flush when you have more solids to dispose of. That can equal water savings that add up to thousands of gallons in just one year of use.

Depending on the model, there may be a pair of buttons on the top of the tank or a lever that can be pushed fully down or just halfway. In either case, you really don't have to change your usual bathroom routine at all. You'll figure out quickly enough when you'll need the full flush and when you can get away with the mini.

A dual-flush will cost more than a conventional toilet but the costs are coming down quickly as more and more people are taking the eco-conscious route. Considering how much water you're going to save, it soon balances out. The installation shouldn't be any different than with a standard toilet, so there are no added expenses there.

This is a better choice for most people than the infamous low-flow toilet. While they do save water, they don't always get the job done. Giving yourself the option to choose your flush, means more efficiency. Of course, this all depends on you taking the time to make that choice. Once you remember the difference between the number 1 and number 2 flush options, you'll be saving water in no time.

HEAT WITH WOOD

GUILT FREE FACT

You don't need to have hardy pioneer blood in you to master a wood stove. If you give it a try, you'll be saving huge amounts of non-renewable fossil fuels every winter.

This is an oldie but a goodie. It's a classic tip that can have huge mileage in saving energy in your home, not to mention reducing your heating bills dramatically too.

Before you get shopping for a woodstove, be advised that it's not as simple as buying a stove at a home improvement store and drilling a hole in your wall for the chimney. Depending on the building codes where you live, you will need an appropriate and suitable hearth under the stove along with certain distances between the stove itself and any nearby walls, and the chimney must also meet proper codes for position and the number of bends in the pipe. Talk to a licensed installer before you shop.

Technical placement details aside, heating with wood is an excellent way to get away from using up fossil fuels. Cutting down trees isn't the most eco-friendly thing to do, but at least the trees can be replanted and regrown. Fuel oil and natural gas are both nonrenewable. They're also a lot more expensive than firewood. You could also be heating with electricity, which also uses up a lot of fossil fuels at the source, unless you're lucky enough to be on a system that uses hydropower.

Wood costs vary widely by region so I can't give you any real figures on what to expect. I personally buy 3 cords of wood each year, and it costs around $850. I would easily spend double that in oil, which I use as a back-up for low heat through cold

Canadian nights. There is no denying that it can take a special skill to start a good fire and to keep it going all through the day, but the basics aren't that hard to master.

You'll need a supply of newspaper and small pieces of split wood (tinder), and a few cords of good wood. Make a pile of small pieces of wood and crumpled paper. Light it, and once that's burning, slowly add larger pieces. You can't start off with a huge log and hope it burns. It won't.

About that wood: hard wood burns hotter but soft wood catches faster. Having a mix is handy though a novice isn't going to know one from the other. Still, when shopping around for firewood, keep it in mind. Hardwood may cost a bit more and that's because you get more heat for your buck.

Plan to Fan

This is one important point when starting to rely on wood heat. Unlike a furnace with a fan or blower, your wood stove just sits there and radiates. That means that your heat doesn't travel around your house in the same way. A ceiling fan helps get air moving, and a strategically placed floor fan can get the warmth down hallways. You can even get small fans that sit right on the stove, and use the stove's own heat for power. They're very neat and won't require any additional power usage to get the air moving.

HAVE SOME HEMP

More than 50.000 products can be made from hemp

Paper pulp

1 acre of hemp = 4 acres of trees

GUILT FREE FACT

Hemp grown as marijuana contains around 3% THC (the psychoactive component), but hemp grown as hemp is far less at 0.3%. Farmers growing hemp are not growing drugs.

There is more to hemp than the usual marijuana debates, and it's actually a very eco-friendly product that needs to be utilized more. It grows fast, requires no pesticides and needs little water. Compared to many other mainstream crops, it's a dream come true for the planet. Let's sum up its greatness, shall we:

- It takes only 1 acre of hemp to produce as much paper pulp as 4 acres of trees

- Cotton uses 50 times as much water as hemp to produce cloth fiber

- More than 50,000 products can be made from hemp. Fibers are used for paper, rope, fabric and other similar products. The seeds produce oil for fuel or food products.

Paper made from hemp can be recycled 8 times before the fibers disintegrate. Tree pulp paper only lasts through 3 cycles.

Hemp first came to North America through Port Royal, Canada in 1606. It was a thriving crop in both the USA and Canada until 1938 as marijuana became more criminalized. More than 30 countries grow hemp legally today. The USA is not one of them. France is the largest European producer, where as China leads the world. Out of 100 known agricultural insect pests, only 8 are a concern for hemp growing. Makes it ideal for organic growing.

STREAMLINE
YOUR CLOSET

GUILT FREE TIP

Stay away from fancy fabrics that require dry cleaning. Nobody needs to add more tetrachloroethylene to the world.

*E*xcessive consumption is the root cause behind so much waste production as well as resource use. We talked about this a bit already in terms of upgrading your phone when you don't really need to, and now we're going to take another look at trimming down your "stuff". In particular, your wardrobe.

The manufacture of fabric can be hard on the environment, especially cotton. Unless it's organic, your cotton clothes contribute to huge pesticide sprayings and water usage. Buying clothes made from hemp is a great start, but it's even better to just buy less in the first place.

Do you have clothes that only ever get worn in a single outfit? In other words, a pair of pants that only goes with one top. Or what about that light sweater that is too warm for summer but too chilly for winter and only ever gets used for about 8 days in the spring or fall?

You might be figuring that since these clothes are already there, what's the problem? You do have a point. I'm talking about it mainly to get you aware of it so you can make more savvy purchases in the future though.

Mixing and matching should be your personal clothing mantra. Try to get pieces that go with everything else. Stick with classic styles that aren't going to be dated in a few months (or weeks). Does it fit comfortably? Do you really need it?

Anything else should be donated to your favorite second-hand shop, so someone else can get some use out of these items without having to buy new either. That's the other way your streamlining can help the planet, by allowing other people to avoid buying new clothes.

GOO DEET-FREE

GUILT FREE TIP

Avoid putting repellent on your face (neck is fine), instead spray a hat with a brim and wear that.

*D*o you spend a lot of time outdoors during the summer, and constantly need a good supply of insect repellent to protect against the mosquitoes? It's great that you're enjoying nature but you might need to green up your bug spray.

Most commercial products contain DEET, also known as N,N-Diethyl-meta-tolumide. Not great for you, and even worse for the environment. The worst scenario is when you douse yourself with this stuff, then end up swimming in a lake. It is toxic to many forms of aquatic life, and it's now found as a contaminant in 75% of American water sources. It's time to do without the DEET. There are more natural alternatives, including some you can make yourself.

The main principle is to have an aromatic oil on your skin to deter any insects who are looking for a meal. They lose your scent and leave you alone. It's not complicated and doesn't really need the chemicals for it to work. Here is a good (and simple) recipe you can try. Mix it up in a spray bottle, and you're ready to go.

- 4 oz. distilled water

- 4 oz. witch hazel

- 1/2 tsp glycerin

- About 40 drops of a pure essential oil

Oils that work great for insect repelling: rosemary, clove, lemon, mint (any variety), cinnamon, lavender, or eucalyptus.

Give a spritz to your clothes and exposed skin when going outside. Keep a bottle with you and plan on frequent applications to keep it going. This mix is all-natural but not as strong as the DEET products.

GREEN UP YOUR
DRYER

GUILT FREE TIP

To keep your clothes soft without the added chemical dosage, use vinegar. Don't worry, you won't smell a thing once your clothes are done. Just add 1/2 cup of white vinegar to your rinse cycle.

ow, we already talked about how air-drying your clothes out on a line is the best eco-option, and I still stand by that. But let's face it, a clothes-line just isn't always going to work in everyone's situation, so why not a few more tips for making better choices while still using the electric dryer?

Keep an Eye on the Dry

Are you running your dryer for an hour, knowing that your clothes will be dry when you return? Well, take some time to experiment with shorter times or cycles. Maybe a load of lighter fabrics (like socks or T-shirts) would be done in just half that time. Don't just let it run blindly. Check periodically and stop your dryer as soon as the clothes are actually dry. You might be surprised how much time you'll save.

Wool Dryer Balls

Wool dryer balls are all the rage these days, and they are a simple little tool you can use to speed up your drying time. They look like large tennis balls made from felted wool fiber. Toss in 3 of them in with your load of wet clothes, and they help to improve airflow through all the fabric. It might sound a little silly but people swear by them. It can shave 5 to 10 minutes off of each load of laundry.

Just make sure you let them dry out themselves between uses. Another plus is that they help soften your clothes and cut down on that nasty static. You probably won't need any dryer sheets once you start using wool balls. If you'd prefer not to pay for a set, you can make your own if you have a supply of wool yarn (must be real wool, nothing synthetic). Just note that buying real wool yarn for this may end up being more costly than just buying the balls, so watch the price tags.

Wrap a tight ball of yarn, making sure to tuck the loose end in. Knot each ball into a piece of pantyhose, and wash with your usual clothes. Dry on your hottest setting. The loose wool fibers should felt together after one or two treatments, creating a solid ball. There, you now have your own wool dryer balls.

KNOW YOUR CODES

GUILT FREE TIP

If your area doesn't recycle 3s, try not to buy anything packaged in PVC.

The recycling codes, that is. If you want to be as effective as possible with your home recycling habits, get to know these numbers. Glass, paper and metal are easy. Plastics are complicated because there are several different kinds, and they're not all recyclable. Most recycling guides tend to describe accepted items by type. Like soda bottles, margarine tubs, yogurt containers and such. That's fine except when you have packaging items that don't match any specific thing on the list. That's where the numbers come in. They are stamped or embossed on most plastics as a number inside the familiar triangle of arrows.

- 1 (PET plastic) - soft drink bottles, peanut butter jars, some kinds of microwave food trays

- 2 (HDPE plastic) - milk & juice jugs, detergent bottles

- 3 (vinyl and PVC) - shampoo bottles, many kinds of clamshell packages, deli wraps

- 4 (LDPE) - some squeeze bottles, most kinds of plastic bags

- 5 (PP) - margarine and yogurt tubs, many kinds of plastic bottles

- 6 (PS) - meat trays, foam packing peanuts, CD cases

- 7 (other) - plastic other than the first 6

Don't just assume that the code means "recyclable". It doesn't. The code just lets you know the kind of plastic you're dealing with. Your own regional recycling program may not accept all 7 kinds. I recently lived in an area where we couldn't recycle 3s or 7s, but I've just moved and our local program now takes all 7. I'm looking forward to diverting more trash to recycling.

You can also use this knowledge to change how you buy products in the first place. Check the code number while still at the store, and try to make your choices so you know your packaging ends up in the blue box instead of the garbage.

SOLID LOTION BAR

GUILT FREE FACT

Sodium lauryl sulfate and sodium laureth sulfate are two other unpleasant additives to many lotions.

A crafty tip today, to help you get away from the chemical-laden hand lotions on the market. Even those that brag about natural ingredients like shea or cocoa butter still have many components that you should think twice about putting on your skin. Various parabens (likely carcinogens), mineral oil, petroleum by-products, artificial colors and fragrances. Why not pamper your hands with something a little less threatening?

Why solid bars? Well, they're a little easier to make than a liquid lotion when you're not using the artificial thickeners and petroleum ingredients. You also don't need to have bottles or jars to store your finished product. Anyway, I figured this was a good place to start as any. To use them, just rub the bar in your hands, sort of like briefly lathering up a bar of soap. The heat of your hands softens up the bar and leaves just enough lotion on your fingers.

Your local health food store/craft shop should have your supplies:

- 1/2 cup coconut oil

- 1/2 cup shea butter or cocoa butter

- 1/2 cup beeswax

If you're not familiar with coconut oil, you can get it at the supermarket but it's a solid at room temperature. Look for a jar rather than a bottle. And the beeswax will work easier if you get small pellets or shaved wax rather than a big block.

Combine everything in a double boiler, knowing that the inner container will likely not be used for food again. A large glass jar will work well, in an inch or two of simmering water. Slowly melt everything together and keep mixing so it's all smooth.

Pour your liquid lotion into your molds. Muffin cups work well, or you could use candy molds. If you half-fill a muffin cup, this recipe will make you about 6 lotion bars. Once poured, let it cool completely, pop your bars out of their molds and you're ready to moisturize naturally.

HEALTHY HERBAL TEA

GUILT FREE TIP

For those smaller over-the-counter issues, reach for a warm cup of tea instead of the latest pill.

*P*harmaceutical options for our minor ailments are common, and that's led to an unfortunate dependence on chemicals that may not be that great for us in the longer run. If you want a more natural option for minor health-related issues, look to herbal teas. Before medicine was available in pill form, this was the usual choice for generations before us. Why not keep the tradition alive? Herbal teas are easy to find, either as a pre-made tea bag product, or as loose dried herbs. Tea bags are simple enough, just steep in hot water for 5 to 10 minutes. If you have loose herbs, you'll need a strainer or a tea ball so you can steep the leaves (or flowers) the same way, then filter out the bits. Either way works well.

Here are some common ailments that can be helped with a simple cup of tea:

- Headache: chamomile, skullcap, ginger

- Upset stomach: peppermint, ginger, catnip

- Sleep aid: chamomile, vervain, valerian (not the most pleasant tasting, but works well with honey)

- Toxin cleansers: dandelion

- Immune boost: rosehip, echinacea

- Reducing inflammation: turmeric, ginger, green tea

EMBRACE THE E-DOC

GUILT FREE TIP

Get a scanner for physical signatures, and then email the file back. Faxing just creates more paper at the other end.

We're talking electronic documents here, not medical advice. By using more and more digital documents, you can use less paper in your day-to-day life. You don't have to be a computer genius anymore to share files and limit your printouts.

Sign up for E-Statements

This the first place to embrace the e-doc, with your banks and utilities. You probably don't really need the paper bills and statements they mail out each month, usually sent along with additional inserts and flyers. You can cut down on a lot of paper use by getting these regular documents via email instead. And to be clear, that doesn't mean you have to pay online. If you usually take your paper bill to the bank for payment, you can just print it out and still do that. You're still saving the paper with the envelope and inserts.

Learn About Doc Sharing

Various services online allow for free document storage and sharing to make it easy to get your papers to the right people, but without the paper. Google has a couple, such as Google Drive (more for sharing existing documents) and Google Docs (more for creating new documents to share). Dropbox is another popular one. You probably won't use them too much in your personal life, but they are excellent tools for the workplace. Stop printing, faxing, and re-printing. Share your docs with these services and it makes life easier as well as paperless.

File Attachments

If you are just passing pages on to one other person, they can simply be attached to an email. There is usually a button that looks like a paperclip in your email program, and you can clip any file to your email and send it off.

START COMPOSTING

GUILT FREE TIP

If your city has no composting projects, drop a line to your local officials and make a polite request.

If you are a regular green guru, you probably already compost all your kitchen scraps. Even so, it's a classic tip that should be followed by as many people as possible. So much organic material ends up in our landfills, and it's actually a bad place for it to be. When mixed in with the "real" garbage, it doesn't have a chance to properly break down and decompose the way it's supposed to. That means it just stays there, taking up space, leading to full landfills much faster than necessary. Keep your organic waste out.

Not only does it mean less bulk for the dump, you will create some rich natural fertilizer if you compost properly. Natural fertilizer like this is excellent for giving nutrients to your plants, and it would mean you can avoid the unnatural chemical products. And it's free. It's a real win-win-win situation.

A bin, box or pile will work fine in just about any size yard. Browse around on Pinterest and you can probably find a hundred different DIY options, or you can just buy a compost bin from the hardware store. Save up your kitchen scraps (avoiding meat and dairy), yard waste, used paper towels, and shredded paper for compost. Keep it stirred if you can, and damp when the weather is really dry. After about a season, you should start to see it break down into crumbly brown "dirt". Shovel that out and give your garden a nutrient boost.

Larger cities are starting to get on the composting bandwagon too, even starting curbside pickup of your organic material for larger composting projects. Toronto, San Francisco, Portland, Seattle and Ottawa are a few examples. It's still a new idea and just picking up speed.

Vermicomposting

For small spaces (like apartments), you can still experiment with composting by harnessing the power of the worm. It's not as gross as it might sound. A large bin filled with wet shredded newspaper makes a perfect worm environment. Get worms from any store that sells "red wigglers" as fishing bait. They will happily live in your bin, and you feed them your food scraps. This approach is better for small-scale composting because you're not relying on random worms and insects to break down your scraps. It's much quicker. Just clean out the newspaper periodically to collect the "castings" (polite name for worm poop) that build up on the bottom. Check out this great online primer to get you started (http://journeytoforever.org/compost_worm.html)

CHOOSING A REFILLABLE WATER BOTTLE

GUILT FREE TIP

Depending on how much water you drink, having a refillable bottle can replace hundreds (thousands?) of plastic bottles every year.

We all know that using those single-serve bottles of water is a terrible choice, ecologically speaking. On average, the USA purchases 50 billion bottles of water, with only 20% getting recycled. Even if you recycle them, avoiding the plastic altogether is even better. Get a good-quality portable water bottle that you can fill for yourself. There are plenty of options on the market, so which kind is the best. There is no single bottle that you must have. It's more about knowing the important features and choosing one to suit your lifestyle the best. Here are some of the things you can look for:

The Right Material

A stainless steel bottle is an excellent eco-choice, as it's very durable and it avoids plastic completely. Glass has the same bonus to it, but is heavy and breakable. For a situation where is just sits on your work desk all day, that may be OK. If you really want or need something made with plastic, get a good quality one that is BPA-free and phthalate-free.

Built-In Filter

This is a great option if you like filtered water on-the-go. If you're already filling up your bottle from a filtered source, then you can skip this type of model. Brita has a few models you can start with to clean up your water on the go.

Insulated?

An insulated bottle usually has a dual layer design to help keep your water cool. For anyone spending a lot of their time outdoors, this can be a pretty nice feature. Like a thermos, they would also keep warm liquids warm if that's your preference (maybe for mobile coffee).

What's on Top?

The top or lid should be considered too. Flip-top or sport nozzle bottles are fine for quick access but the pieces are usually less sturdy and are more likely to leak of your bottle gets tossed around a lot. A screw-top is more durable as long as you keep track of the lid once you take it off.

GET RID OF HOUSEHOLD PAPER

GUILT FREE TIP

If you're shopping for a handkerchief, try the men's department around Father's Day. They're a popular gift.

We all know that disposable products are lousy things to have around the house, as they waste resources in their manufacture and build up in landfills after you trash them. As an eco-conscious person, you already know this and have likely gotten rid of all the plastic cutlery in your kitchen and those single-use dusting pads. But it can be much trickier to get rid some of the usual household paper products. Kleenex and paper towels are what I'm talking about.

They tend to be lost last holdouts in an otherwise green kitchen, mainly because they are involved in icky messes that are so much simpler to dispose of. Well, now it's time to think about it.

Paper Towels

A collection of light cotton dish towels can go a long way to reducing wasted paper in the kitchen. They're fine for quick wipes and mop-ups, and then you just toss them in the wash. Leaving damp or very soiled towels lying around for too long can be unsanitary, so expect a little more laundry. Don't let the slight increase in water usage bother you. You're still making the greener choice by cutting down on your paper consumption. It takes 17 trees and 20,000 gallons of water to produce 1 ton of paper towels (cite below), so you can see that there are a lot of resources at stake beyond your laundry water use.

We seldom use paper towels at our house, and I have about 10 dish towels in the laundry each week. That's not even half a load. While we go with reusable cloth most times, certain really offensive messes get the paper towel treatment.

Tissues

Kleenex is another spot where it can be hard to visualize a reusable option. But remember that carrying a handkerchief used to be the norm, for both the gentlemen and the ladies. Either buy a few, or save the money by making your own with basic soft quilters cotton.

Worried about carrying around a pocket-full of germs? As long as you toss your hanky in the wash each day, there should be no worries. You can also stick with disposable tissues when you are particularly sick, and the hanky for the rest of the time.

HOW CLEAN IS CLEAN?

GUILT FREE TIP

Skipping one shower will save between 15 and 20 gallons of water. If you dropped one or two showers each week, that's a lot of water savings over the months.

This is a bit more of an idea to reflect on instead of a concrete tip. How clean do things really need to be? Do you really need to throw that T-shirt in the wash after wearing it for a single non-sweaty afternoon? What about those jeans that never saw any actual dirt?

In our typically affluent North American world, we have developed a biased attitude towards cleanliness as well as the constant need for new things. Anything slightly used is dirty and therefore unacceptable. It's time to get out of that rut.

I'm not saying that you have to walk around with spaghetti sauce on your shirt or grass stains on your knees. It's about being reasonable, allowing a little wear before immediately adding something to the laundry. You don't have to go back to your college days, where you give everything the sniff test to find something presentable for public wear.

By allowing a few items to last more than one wear, you can cut down on your weekly laundry by a full load or even more. Saves time for you, as well as a considerable amount of water and energy.

And while we're talking about acceptable levels of cleanliness, the same ideas can be used in your showering routines. Are you a daily showerer? Do you really need to be?

This is a big hot-button issue with many people, green-minded or not. The idea that skipping your shower for one day makes you dirty and smelly is not accurate for many people. Of course, if you do a lot of physical labor or the weather is hot, then it's going to be necessary.

But are you really dirty after a lazy day around the house, or a slow day sitting at an office desk?

LED
LIGHTS

GUILT FREE TIP

If you're trying to go green with your light bulbs, it's time to move past the CFLs and try some LEDs.

The humble light bulb got a major green overhaul several years ago when the compact fluorescent bulb was first introduced (aka the CFL). Many people embraced them because they use a fraction of the energy a classic incandescent bulb does, and they last for years. Well, years have passed since the CFL revolution and you should get up-to-date on the latest in green lighting.

The newest choice for environmental light bulbs are the LEDs (light emitting diodes), and they are just starting to get commonly available in household bulb form. They used to be only available as small task lights or strings of Christmas bulbs. Now you can get LEDs to replace your everyday lamp bulbs. Here is a comparison between equivalent incandescent, CFL and LED bulbs so you can see the differences:

	How much energy used	How long they last
Incandescent	60 watts	1,200 hours
CFL	14 watts	8,000 hours
LED	7 watts	50,000 hours

One of the long-standing problems with CFLs is the mercury content, and that many won't work well in cold weather. Both problems are resolved with LEDs. Add in the even more reduced energy consumption and the fact that they can easily last 5 to 10 years, they are a really green choice.

Just be prepared for a sticker shock. We were replacing some bulbs in our new home, and were a bit put off by the $20 cost for a single LED bulb. Given the teeny-tiny level of power consumption, we figured it was worth it in the bigger picture.

MAKING HOMEMADE TOOTHPASTE

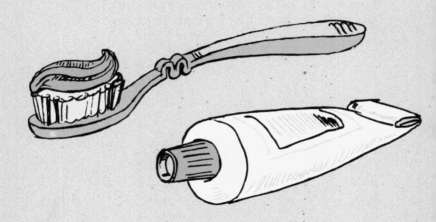

GUILT FREE TIP

If you don't like the communal toothpaste jar idea, you can make up smaller batches so everyone in the family has their own.

Time to talk about toothpaste. It seems harmless enough, until you notice that it says right on the tube that you're not supposed to swallow it. That's because it's not actually good for you. Sodium fluoride, sodium lauryl sulfate, and triclosan are all very questionable ingredients that you shouldn't really be putting in your mouth on a daily basis. Not only are they not great for your body, they end up in the water supply every time you spit after brushing.

There are a few natural toothpastes on the market but you can also make a batch of your own for a fraction of the cost. All you need is a mild abrasive and something to make it paste-like. A good basic recipe includes:

- 1/2 cup coconut oil

- 2 tbsp. baking soda

- 15 or more drops of mint oil (or any flavor you wish)

- 1/2 tsp stevia powder (optional for sweetening)

Soften up the coconut oil until its smooth enough to stir around. Then mix all the ingredients together until they are evenly blended. It should stay as a nice paste at room temperature so you can keep it in the bathroom. Just dip the bristles of your brush in and use the same amount you would a commercial toothpaste.

GO VEGETERIAN

SWEET

SUGAR

GUILT FREE TIP

Even if you don't choose to go all the way, try to incorporate more plant-based foods into your diet and cut back on the meat. Every little bit helps.

This isn't a simple or easy tip, and it might be too drastic for your family. That's fair. Even a slight reduction in your meat consumption can make a difference though. Keep reading to find out more.

Many people think that giving up meat (either vegan or vegetarian) is an issue that's all about animal welfare. There is a lot more to it than that, and it can be a choice that benefits the whole planet. The industrial raising of livestock for meat is an incredibly wasteful industry, and is unnecessarily draining on the environment. Raising plants for food takes up less space, fewer resources and creates less pollution.

To grow and process a pound of wheat, it will take 25 gallons of water overall. For a pound of beef, the entire process will devour more than 2,000 gallons. Not only does the animal have to drink, but there are still crops grown for animal feed. Clean water isn't in abundance like it once was, so why make a choice that wastes so much of it?

And what about all the space? Not only are huge tracts of land needed for pasture or grazing, there are further fields needed to grow the various crops (corn, soy) to be used as animal feed. Approximately one third of the entire planet's surface is used to raise animals for meat consumption (cite below). And it continues to grow, with thousands of acres of rainforest being cut down daily to make more room for grazing cattle.

Some people say that greatest change you can make in your life to help the environment is this one.

ARE WE RECYCLING ENOUGH?

GUILT FREE TIP

More than 60% of the average household waste is compostable, yet only 8% is composted.

With the birth of the handy blue bin recycling program, it has become easy to separate our trash so that recyclables can be reclaimed. So now, everyone recycles, right? Unfortunately, that's not even close to the reality. According to the EPA, the USA produced 254 million tons of garbage in 2013. Approximately 87 million tons were diverted to recycling or composting.

That's only a 34% recycling rate. People recycle newspapers the most, at 67% and glass was one of the most seldom recycled coming in at only 5%. Even the ubiquitous plastic was only reclaimed at 9%. Only 25% of the growing "electronic waste" is getting recycled

The "blue box" program originated in Canada. Similar curbside programs are now in place around the world, with or without the literal blue box. There are approximately 8,000 curbside recycling programs in the USA.

A GREENER HOUSE PAINT

GUILT FREE TIP

You can avoid paint completely by choosing wallpaper, fabric or wood paneling for your walls. Consider all your options.

*P*ainting the walls in your home is one of the easiest and most common ways to revamp or restyle a room. But you'll also be layering on heavy toxins if you're not particular about the kind of paint you're using. The air quality in your house is very important, and it should be a clean and healthy place.

Standard house paint is loaded with chemicals known as volatile organic compounds (or VOCs), used to make the paint have a smooth finish, to dry faster and to prohibit any mildew growth. That's all well and good, but the fumes given off into the room are horrible for you. You already know how you can get a headache or nausea if you leave all the windows closed while paint, so none of this is likely a surprise.

Most paints are going to contain varying levels of VOCs. With a little research and persistence, you can hopefully find some more environmentally friendly options. Several major brands have at least one line of paint either zero-VOC or low-VOC (that's how they'll be labeled), but it will likely give you a limited set of options.

You can also go old school with some vintage style paints known as "milk paint". More commonly used for furniture or craft items, you can do room walls with it too. Colors tend to be more muted and they don't have a glossy finish. These types of paint are actually made with milk protein as a binder, making it natural and far less toxic than other modern paints.

Most milk paint comes sold as a powder, which you mix up yourself. It's really simple, and you just mix up a batch right before you paint. They dry pretty quickly and you won't need to air out the whole house for days to keep the air breathable.

Don't automatically plan for 2 coats either, and that goes even if you are sticking with conventional paints. People tend to use a lot more paint than necessary when the second coat is barely even noticeable.

THE REALITY OF THE ELECTRIC CAR

GUILT FREE TIP

Mid-distance cars like the Smart Car or Nissan Leaf can be yours for less than $30,000 right now.

I thought we'd take a broader look an up-and-coming green option that is outside the house for a change. The electric car has long been seen as a bit of a "holy grail" in the world of eco-friendly pursuits, as it would remove the need for fossil fuel as well as a major course of pollution. An electric car is clean, quiet and if you recharge it from a renewable energy source, it will use up no resources to run. How perfect is that? So, how close are we to the era of the electric car? It's hard to say. Here are some of the positive improvements in the industry that are slowly but surely bringing the electric car to the masses.

They're Going Farther

It used to be that you could only go to the local mall and back before you had to charge up your car. Today, there are electric car models that will go anywhere from 60 miles to more than 100 miles between charges. Ok, still not great for the Great American Road Trip, but definitely reasonable for a daily commute-to-work vehicle.

There are More Charging Stations

If you are limited to charging your car at home, it's going to be very restrictive. Technology is rapidly changing and investments are being made in a variety of public charging stations across the country. There are currently more than 20,000 in the USA and the number is growing. Depending on the style, you may have to park and let your car charge for a few hours (it's still not like filling up a tank with gas). Visit www.plugshare.com for a map and current status of charging points across North America.

They're Getting Cheaper

This is probably the most important point. The cost of electric cars is finally getting low enough to make them a reasonable alternative for the average person. If you do have a higher budget, you can spring for a snazzy Tesla Model S (over $60,000). Sure, it's not going not going to be affordable for everyone at this price tag but it's a lot closer to mainstream prices than ever.

I think that we're going to see electric vehicles become more common within the next 5 to 10 years, as costs come down and charging access goes up. Once it reaches that vital tipping point, those old gas-guzzling cars will be a thing of the past.

Mid-range cars like the Smart Car or Nissan Leaf can be yours for less than $30,000 right now.

UPCYCLING OLD CLOTHES

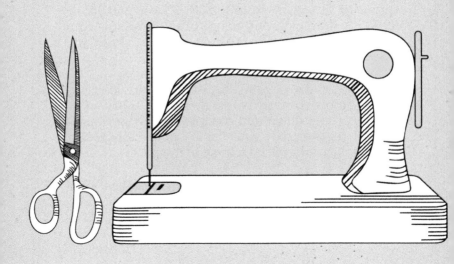

GUILT FREE TIP

You can cut your clothing down into dish cloths and dusting rags.

U pcycling is the new, trendy form of recycling where you take old stuff and remake it into something better. Right now, let's talk about old clothes and what you can do with them.

Firstly, any clothing that is done at your house because it no longer fits but it is otherwise fine for someone else should be given to your local second-hand shop or Freecycled to someone who can use it. Clothes with wear in them should be worn. But anything that is stained, damaged or otherwise not suitable for wearing anymore shouldn't go in the trash. There is still a lot of use in that holey shirt or threadbare jeans.

Salvage any useable notions, like buttons or snaps. You never know when an odd button can come in handy when fixing up another piece of clothing. Once you've scavenged your item, think about the fabric. This is where you can really clean up, materially speaking. Slice out the damaged areas, and seam lines. Try to keep the chunks as big as possible and you can have a lot of material to make something new. A simple quilt is the most common project that can use all kinds of mismatched pieces, even the smallest snippet of material.

If you're handier, try making new clothes or clothes for kids. A large man's shirt can yield enough material to make something for a child for example. Smaller projects could be coasters, or patchwork designs in wall hangings, curtains or a custom tablecloth.

Denim from jeans is extra-sturdy and can be used for clothes or quilts, especially any project that needs the tougher fabric for outdoor use. I have a stack of jean legs pinned together to become a huge outdoor picnic "blanket" as soon as I get my sewing machine out of storage. A simple quilt is probably the easiest for any non-sewers out there, but if you have a few decent skills under your belt, you can really make a lot of things for next to no money. Not only are you helping your budget bottom-line, you're keeping bags of fabric out of the landfill.

TRY PAPERLESS
GIFT-GIVING

GUILT FREE TIP

During Christmas, there is more than 25 million tons of garbage created just from wrapping paper and related holiday packaging. That's a lot of unneeded waste.

*E*ver notice how you create a huge pile of discarded paper after the gift-giving frenzy of Christmas or birthdays? Sure, this might only happen a couple times a year, but it does mean a lot of garbage is being produced and a lot of trees are cut down to make that paper in the first place. It does serve a purpose, except it is a very temporary purpose and one that could be better filled with other products. Next holiday time, try to go paper-free. Or at least, cut down on the paper that ends up in the trash.

The simplest approach is to switch from wrapping with sheets of paper to using gift bags. This is our tradition here at my house. I have a big tote box filled with bags for all occasions in a mix of sizes. Not only does it make wrapping faster and cheaper, it means no paper is wasted with each gift. Item goes in the bag and the top is stuffed with colored tissue paper (also saved each year). Younger kids might not be able to resist the temptation of peeking through the tissue, so you might have to tape or staple the bags shut. That's your call.

Fabric bags are another option, and they usually tie shut for better gift security. With a little creative folding, you can use fabric in place of paper without resorting to using bags. Ribbon or pins can hold it all together, and a fabric-wrapped gift will look very pretty under the tree. Maybe use some of that reclaimed material I just mentioned in the last tip?

Another way of looking at the problem is to reduce the number of wrappable things being given to begin with. Gift cards don't require any wrapping, nor does cash. Of course, you don't want to reduce all you're giving to a few tiny cards. Your Christmas tree is going to be pretty bare if you do. It's just one thing you can do along with the others to reduce your wasted paper.

If you do need to resort to paper, try to get some that is already made from recycled material or some that can be recycled itself. Some of the ones with heavy gloss or metallic finishes cannot be recycled, making them doubly wasteful.

Hit Pinterest and see what crafty options you can find for making big frilly bows out of strips of newspaper or old fabric. Your presents don't have to be plain to be green. Even glossy junk mail can be cut down to make colorful bows if you get tired of the black and white look.

GUILT FREE FACT

Traces of the chemical ingredients end up going down the drain eventually, and they are in our water supply.

Whether you call it soda, pop or just Coke, these heavily sweetened drinks should be dropped from your daily routine. Firstly, the stuff is terrible for your health and being eco-aware means also taking care of your body as well as the planet. Secondly, there really is a larger impact on the environment when it comes to the soda industry.

It can take more than 100 gallons of water to produce a 2-liter bottle of soda (cite below), though that varies by brand and flavor. Ingredient-wise, you have to deal with high fructose corn syrup, brominated vegetable oil and lots of toxic "sweeteners" if you insist on drinking diet. Don't forget the containers. Plastic pop bottles and cans make up a large portion of our recycling, and even though some (not all) are reclaimed, that's a lot of plastic manufacturing going on.

Fruit-Infused Water

Fruit flavored water is a much better choice over pop, and you can make it yourself. There really isn't much of a recipe or technique to learn. It's simply a mix of water and a collection of sliced fruit, vegetables or herbs. Let it chill and you're done. To get you started, here is a refreshing recipe:

- 1 cup strawberries, sliced up

- 1 cup cucumbers, sliced up

- 1 lime, sliced

- 1/2 cup fresh mint, torn up

- Water and ice cubes

Starting with a 2-quart jug, pile in the ingredients along with a couple handfuls of ice cubes. Fill up the rest of the pitcher with water and let it chill in the fridge for at least half an hour. Serve it up as soon as you're thirsty after that. And to go back to the first point for a moment, don't forget that things that are bad for your body often still end up in the environment.

A BETTER EARTH, ONE CAP AT A TIME

GUILT FREE TIP

Plantations that produce shade-grown coffee don't raze the land. Instead, they plant their coffee among the trees and maintain the ecosystem much better.

Coffee may not seem like an environmental concern, especially when you're enjoying that first morning cup and nothing else matters. Are you making green choices when you it comes to your coffee?

The Pod Controversy

Are you a single-serve coffee brewer using one of the latest machines that take pods? While we all love the convenience and freedom that a pod system can bring to our mornings, they are an ecological disaster. Pods are usually not recyclable and they are producing a huge amount of plastic waste. To be fair, many of the major manufacturers of these things (like Keurig and Tassimo) are partnering with recycling organizations like Terracycle to maintain drop-off points where you can dump your empty pods for recycling. Better than nothing, but it's still a pain to do.

There are some third-party refillable pods out there that you could also try, using your own ground coffee. You just wash and reuse the pod in most cases.

Shade Grown Coffee

We all know about fair-trade coffee that is produced and sold in a way that fairly pays and treats the workers (coffee is often grown in third-world areas where workers are otherwise paid poorly or exploited). But on the environmental front, you want to look for shade-grown coffee as well.

Mobile Coffee

Are you getting your coffee while on-the-go? Those disposable cups are going to start adding up, if you do this week after week. Try making coffee at home and using a thermal mug to take it with you. Perhaps your local coffee shop would pour you a cup in your own mug? If you do have to take a disposable, make sure you toss it in the proper recycling bin as soon as you are done.

Compost

One last green thought. When you make coffee at home, your used grounds are ideal for the compost bin. Don't trash them. If you use a coffee machine that takes paper filters, you can just take the entire thing, paper and all, and dump in the compost. The paper breaks down just fine.

RESPECT HAZARDOUS WASTE

GUILT FREE TIP

Take your unwanted products (both prescription and over-the-counter) to the pharmacy and see if they can dispose of them properly. Don't trash or flush any meds.

The term "hazardous waste" might make you think about radioactive plutonium or something, but there is almost always some hazardous substances in every home and you should take care not to let them end up in your garbage.

Paint

Don't pour any leftover paint down the drain, even if it's only a few drops. Paint contains harsh chemicals that water treatment plants aren't designed to handle. Check with your local waste disposal company for exact details, though most recycling programs will take paint cans as long as the residue of paint in them is dried. So you just need to leave a can open for a few days to dry out the bottom. If there is more than a half-inch or so of liquid, then you'll need to find the proper place to take it. A hazardous waste depot should take liquid paint.

Batteries

I know it's easy to toss that little battery in the trash, but please don't. Batteries are loaded with heavy metals, acid and a disturbing mix of other toxins. Ideally, you'll be using rechargables so that fewer need to be disposed of any way, but there are always going to be a few things that have one-time use batteries. Most hardware or electronics stores offer drop-boxes for batteries, whether you have a handful or just one. On average, Americans buy more than 3 billion household batteries each year (cite). Let's not have all those end up in the trash.

E-Waste

This can be old computer speakers, an old phone, TV remote, or any other electronic device that has no more life left. They are filled with heavy metals and plastics and should not be thrown away. All of these devices can be recycled to reclaim a lot of their materials, and most electronics stores offer drop-off points.

Medication

This is one many people forget to think about. Unused or expired medication shouldn't be thrown away either. These products add so many chemicals to the landfill or water supply that they are just as hazardous as batteries. According to the EPA, more than 100 different medical products are found in various samples of drinking water around the country (cite).

BEWARE THE GREENWASHING

ORGANIC PRODUCT

100% NATURAL

GUILT FREE TIP

Be aware that labels lie all the time and companies are constantly jumping on the living green bandwagon to attract new customers.

*G*et overwhelmed by all the claims you see on products, telling you how green/natural/healthy they are? It can make it hard to make sensible choices because none of these terms really means anything. Manufacturers slap them on all kinds of things without having to back up their claims. We call this "greenwashing". They're trying to whitewash over their content with misleading green claims, hence the new term of green-washing.

Your first step is to look for the organic label. Out of all the terms you might find on a product, that's the only one that is federally regulated and it has an actual meaning. It means all the ingredients were grown without chemicals (you can read all the details on the organic label in an earlier tip).

After that, take the time to read the ingredients. That's the only way to really know what's in something. A label for hand lotion may brightly proclaim that it's made with all-natural cocoa butter, but that's probably not the whole story. It likely also has a mix of parabens and sodium laureth sulfate. Not a great choice, and certainly not the "all-natural" as it claims.

ARE CHEMICALS BAD?

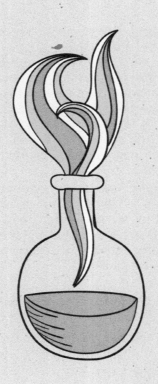

GUILT FREE TIP

You need to be aware of what you're eating as long as you don't get overly paranoid about things that are actually very natural.

$\int$ ince we were talking about reading ingredient lists to avoid greenwashing just a moment ago, I thought it was time to introduce some more info on the chemicals we see in so many products. The general attitude is that we should avoid "chemicals" and go for natural ingredients. It's a fine approach, except it's too simplistic. A lot of chemicals exist harmlessly in nature and are no problem at all, until you read their fancy chemistry names and get intimidated.

A good example is water. No one would question that water is a fine, natural and beneficial ingredient. But what if you read "dihydrogen monoxide"? Don't get fooled by chemical names that really represent natural or otherwise harmless compounds. Things like:

- Acetic acid - main component in vinegar

- Lecithin - a fatty substance that comes from eggs

- Ascorbic acid - vitamin C

- Tocopherol - vitamin E

- Pyridoxine - vitamin B6

- Fructose - naturally occurring sugar in fruit

- Glucose - naturally occurring sugar in fruits, vegetables and starches

This isn't a complete or comprehensive list, just a place to get you started and to make the point about chemicals. Everything is a chemical, whether it has a scary name or not.

THE PLASTIC BAG BATTLE

GUILT FREE FACT

Plastics bags can take up to 1,000 years to finally break down.

More and more people have adjusted to carrying their own reusable bags when shopping, and many large cities have gone so far as to ban the use of disposable plastic grocery bags entirely. These are great steps but don't believe that the bag battle is over. There are still far too many of them out there, and they are a major hazard to wildlife and any natural ecosystem when they end up free flying as litter.

Approximately 500 billion to 1 trillion bags are used every year worldwide, and though the majority are reclaimed through recycling or trash, the estimated 2% that are loose litter still equals millions of bags out in the ecosystems.

The average US family brings home 1,500 plastic bags every year.

Estimated 100,000 marine animals and 1 million die every year from plastic litter in ocean water (all sorts, not just bags). They either tangle in it, or eat it accidentally.

One in three leatherback sea turtles has a plastic bag in its stomach. Sea turtles often mistake floating bags for their usual diet of jellyfish.

CLEANING DAY

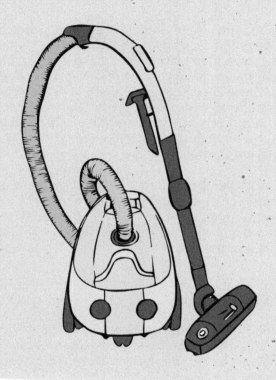

GUILT FREE TIP

For a quick disinfectant, add half a cup of vinegar to your homemade cleaning solution.

*Y*ou've probably collected a number of green cleaning tips and recipes, and now it's time to pull them all together to create a cleaning routine. Why not take a look at how I get most of our cleaning chores done around my house?

My first concern is usually the weather. To make use of our clothesline, I have to plan on doing laundry on a suitable day to line dry. That means sunny and hopefully with a light breeze, but I can work with any day that's not raining. We use a natural laundry soap with no fragrances. I hope to try making my own soon once I get a decent source for buying washing soda.

Once that's underway, I tackle dirty surfaces. I have a spray bottle of general cleaner that I use for just about everything (the recipe is shared in an earlier tip). Ours is scented with a little extra tea tree oil. A few old cotton rags and my spray bottle is all I need, and no garbage is created while I wipe. Around the sink or tub, I can sometimes use a little extra scrubbing power. I just fill a little dish with baking soda (we keep a large bin of it in the pantry), and dip my rag in that before I scrub. Rinse with a little water, and everything comes clean.

We live on a small farm so I don't really expect my floors to ever be "clean". So mopping isn't a regular chore unless the floors are getting noticeably sticky or muddy. I just use a simple sponge mop and clean water. For periodic disinfecting, I will add a half-cup of vinegar to the bucket.

As you can see, a green routine isn't any harder or time consuming than one that uses excessive chemicals. It takes me the same amount of time to wipe out a grubby sink with my vinegar spray than it would with one filled with ammonia or bleach.

My last regular chore is the toilet. I keep it really simple, and pour in 1/2 cup of baking soda into the bowl, leaving it to soak while I work on another chore. About half an hour later, I give the bowl a scrub with a brush and a rinsing flush.

I also dust once in a while, usually on a different day from the "wet" cleaning just so I don't spend the whole day doing housework. We use rags to dust, no sprays or disposable cloths and it works fine. I enlist my daughter to help with this one so it goes quickly, again creating no waste or garbage.

When we're all done, the house smells fresh and clean, not overly fragrant from a dozen different scented cleaning products.

WATCH FOR
WATER WASTE

GUILT FREE TIP

Adding a low-flow shower head can reduce the amount of water being used while you bathe.

We use water in a lot of different ways around the house, and careful usage can make a big impact on how much we waste. The worst offense is when the tap is just running, and clean drinking water literally disappears down the drain having done nothing for us whatsoever.

Don't discount a few minute of unnecessary tap flow either. It may not seem like much, but the average person wastes about 1,500 gallons per year just in by having the water run while they brush their teeth. When you're busily scrubbing away at your teeth, turn off the water.

If you wash dishes by hand in your house, don't let the water run in the second sink continually while you rinse. Have the second sink partially filled, and dip your dishes into that or use a drying rack instead. Fill it up with soapy dishes and use the faucet to rinse them all at once (a tap aerator works very well for this). On that note, we don't really rinse dishes at all in our home because we use an all-natural dish soap that doesn't leave much detergent residue on our dishes.

Another water run-away in the kitchen can be waiting for the water to be hot (or colder) from the tap. Keep a pitcher of water in the fridge for a cool drink, if it's chillier water you're after. Waiting for warm? It's not as easy to have warm water on hand, so you'll still likely have to let the water run. The catch is that you don't need to let it go to waste. I have two jugs under the sink, and when we have to run the water waiting for it to warm up, the jugs get filled. We use that through the day, and it saves gallons each day from going down the drain.

Showers are another notorious water waster, with nearly 3 gallons disappearing for every minute the water is running. So you can start saving water just by trimming a minute or 2 from your shower. You'll hardly notice that at all. Lose track of time while showering? Have a clock in clear view, or even set a kitchen timer to remind you when your time is up.

GREAT GARBAGE GYRE

GUILT FREE TIP

Pick up a stray water bottle on the street and toss it in your own recycling. Costs you nothing but a moment of your time.

After talking about plastic bag litter and how they are contaminating so much of our oceans and land ecosystems, I thought it was time to mention one of the biggest (literally) trash problems in the world: the Great Pacific Garbage Gyre. Also called the Pacific trash patch, this is a huge collection of free-floating debris that is building up in the ocean.

Unfortunately, the phenomenon is virtually invisible and gets too little attention from mainstream media. The problem is that the garbage building up in the twisting currents of the North Pacific has been broken up into tiny particles that you can't easily see, even if you were to boat through the area. Just because it's small, doesn't mean it's not a terrible threat. These particles get eaten, both by tiny animals that mistake the pieces for food, and by larger animals that simply ingest the plastics by mistake as they are eating their usual larger diet. All this eaten plastic eventually kills the animals and birds.

Estimates say that the patch is as large as the state of Texas, and some say much larger. Millions of animals and birds are killed every year from consuming this soup of plastic bits. And just because this is all going on in the water that it's not an issue involving the litter you see every day. About 80% of the garbage in the gyre has originally come from land trash so it's not all about boat debris.

Cleaning up this monster is going to pose a serious problem because you can't just net out the bits with a big trawling net. They're just too small and the area is just too big. So the best we can do at the moment is to keep it from getting worse. It's not about recycling as it is about loose free-flying litter. Make sure your recycling bin has a lid during windy days and make an effort to pick up any trash you see when you're away from home.

Plastic trash and litter are having a much bigger impact than just making the roadside look messy.

GREEN UP YOUR
FAST FOOD

GUILT FREE TIP

*Consider the impact if everyone stopped taking piles
of napkins or only used one plate at the buffet?*

OK, fast food and environmental awareness don't typically go hand-in-hand but we all end up at a McDonald's now and then out of indulgence or desperation. That doesn't mean you leave your eco-choices at the door.

Take One Napkin Each

Unless you have small kids in your family, you probably don't need more than one napkin per person. Big fistfuls of paper go to waste with each meal because people tend to take 3 or 4 napkins "just in case". Not very messy Carry a small handkerchief with you to wipe those greasy fingertips, and then take it home with you to wash.

Take One Ketchup Cup

If you're eating at one of those places that have a big ketchup pump and little paper cups, just take one. That's right, just one. When you run out of ketchup, go back and fill it up again. I usually go back to the condiment island 3 times and I've survived just fine.

Do you Need Straws?

Unfortunately, most fast food drink cups are flimsy and require the plastic lid to help them hold their shape. And with the lid in place, you don't have a lot of choice with a straw. But if you can take a drink cup without a lid and a straw, that's less plastic being used. A real eco-warrior would bring their own stainless steel straws (yes, that's a real thing) to avoid the extra plastic.

Garbage Right

Many fast food joints offer recycling, usually a generic bin where all plastic, paper and glass are dumped in. Some places are now offering a more refined system where you separate out your recyclables and even compostable food waste. Don't be lazy and dump everything in the "trash". I know you're probably in a hurry but if you have time to recycle at home, you have time when you're out.

Keep your Buffet Plate

Not quite fast food, but a little tip when you're at the next buffet restaurant. The custom is to leave your used plate at the table, and get a fresh one when you go back to the buffet. If you usually go back 2 or 3 times, that makes a lot of dishware to wash. If everyone kept their plate until they were finished, they'd be able to save significantly on water and utility usage.

GET TO KNOW GMO

More than 90% of all the soy, cotton and corn are GMO

94% of cotton in 2015

94% of soybean in 2015

92% of CORN in 2015

GUILT FREE FACT

The Bt toxin is apparently safe for consumption because it easily breaks down during digestion.

When it comes to healthy eating, there is no hotter buzzword than GMO. The problem is that most people don't really know what it means or how "Genetically Modified Organisms" are affecting our food supply. Are these things really in our food or not? Is it as bad as some people make it out to be? Maybe it's worse?

More than 90% of all the soy, cotton and corn planted in the USA is somehow genetically modified. In 2015, the statistics were 94% of cotton, 94% of soybean and 92% of corn.

Sixtyfour nations require labeling of food products containing GMO ingredients, such as Spain, Australia, France, UK, Sweden, Italy and Russia. The USA and Canada do not.

Most common GMO crops: corn, soy, yellow squash, alfalfa, canola, sugar beets.

Organic foods (with the proper certified label) do not contain any GMO ingredients.

Most common modification is known as "Bt", and these crops contain genes from a microbe called Bacillus thuringiensis, that makes them naturally produce their own pesticide.

MAN VS MACHINE

GUILT FREE TIP

Dishwashers can easily fit into an environmentally-conscious household with our tips!

*Y*ou might be picturing the Terminator now but I'm talking about a battle that's a bit more domestic. It's the ongoing debate on whether washing dishes by hand or with a dishwasher is more efficient. There has been a longstanding opinion that dishwashers (the machine kind) are water-guzzling beasts and that it's better to wash your dishes by hand if you want to be environmentally friendly. With today's modern machines with their water-saving cycles, it's getting to be a close call.

It's not easy to compare because it depends on how many dishes you're doing, which cycles you choose for the machine and how you wash dishes by hand. Do you leave the water running in a second sink for rinsing? Do you run the machine half-empty. All things to consider.

The main point that usually tips the scales one way or the other is leaving a tap running to rinse. Over the course of a large load of hand-washed dishes, that really adds up to a lot of water consumption. In that scenario, the dishwasher wins without much doubt. But if you don't leave the tap running, then it's less obvious.

A German study (cite) has declared the dishwasher to be the winner but their study seems to assume a huge amount of water used by the hand-washers. Again, the trick is how you wash your dishes. The average dishwasher uses around 4 gallons per load. If you only fill the sink with wash water, and don't run-rinse, you will almost certainly use less than that.

Each minute of rinsing will use up to an additional 2 gallons of water, and that will add up fast.

If you use a fairly natural dishwashing soap, you shouldn't need to rinse too much. Alternatively, you could add a gallon or two of water into the second sink, and dip to rinse rather than run the tap. With these few tips, hand-washing is the winner over the machine, but it's still close.

AUTHOR BIO

Terri Paajanen

Terri Paajanen has been living on 5 acres for the past eight years, growing a wide mix of fruit, vegetables, herbs and even a few apple trees in her gardens. Before that, she worked hard to produce food in various small city lots, containers and balconies. Over the years, she's honed her backyard agriculture skills and learned the tricks to a successful garden without the heavy use of chemicals or synthetic products. Terri has a B.Sc in Biology and botany and has mastered the fine art of putting bushels of food on the table every year. She is looking to expand her home's acreage and plans to experiment with nut trees, more fruit and joining a new farmer's market. Terri is also the author of 52 Simple Ways to Live Green published in 2014.

Madeleine Somerville

Madeleine Somerville is a writer, blogger, and author of All You Need Is Less: A Guilt-Free Guide to Eco-Friendly Green Living and Stress-Free Simplicity. She lives in Calgary, Canada with her three-year-old daughter and writes at SweetMadeleine.ca.

CPSIA information can be obtained at www.ICGtesting.com
Printed in the USA
BVOW05s2336121016

464891BV00003B/3/P